Long ago there was a good hobgoblin. His name was Lucky. Lucky wanted to live on a ship.

The skipper of the *Atlantic* said to Lucky, "You can sail on my ship any time!"

"That is fantastic," said Lucky.

Lucky worked hard on the *Atlantic*. He would run up the masts. He would fix the sails.

One night the *Atlantic* was lost. But Lucky could see in the fog.

"I see bright lights," said Lucky.

"You are fantastic," said the skipper.

Lucky was even a good cook. His dinners were excellent!

"This food is fantastic," said the skipper.

Lucky had another important job. He looked for big storms.

"This darkness looks endless!" said the hobgoblin. "There will be a great big storm today."

Once Lucky could see an iceberg.

"That iceberg could harm the *Atlantic*," said Lucky. "I must tell the skipper!"

Lucky saved the *Atlantic*.

The ship did not hit the iceberg.

"You are an important friend," said the skipper.

"This is fantastic!" said the happy hobgoblin.

The End

Understanding the Story

Questions are to be read aloud by a teacher or parent.

1. What is the title of this story?

2. What is the hobgoblin's name?

3. Why is Lucky a good name for him?

4. What does Lucky say about getting a medal?

Answers: 1. *A Hobgoblin Saves the Atlantic* 2. Lucky 3. Possible answer: because he saved the *Atlantic* 4. "This is fantastic!"

Saxon Publishers, Inc.
Editorial: Barbara Place, Julie Webster, Grey Allman, Elisha Mayer
Production: Angela Johnson, Carrie Brown, Cristi Henderson

Brown Publishing Network, Inc.
Editorial: Marie Brown, Gale Clifford, Maryann Dobeck
Art/Design: Trelawney Goodell, Camille Venti, Sarah-Beth Zoto
Production: Joseph Hinckley

© Saxon Publishers, Inc., and Lorna Simmons

All rights reserved. No part of this publication may be reproduced, stored in a retrieval system, or transmitted in any form by any means, electronic, mechanical, photocopying, recording, or otherwise, without the prior written permission of the publisher. Address inquiries to Editorial Support Services, Saxon Publishers, Inc., 2600 John Saxon Blvd., Norman, OK 73071.

Printed in the United States of America
ISBN: 1-56577-995-9

Phonetic Concepts Practiced

vc|cvc|cv (Atlantic)

ISBN 1-56577-995-9

Grade 1, Decodable Reader 33
First used in Lesson 92

What Is My Pet?

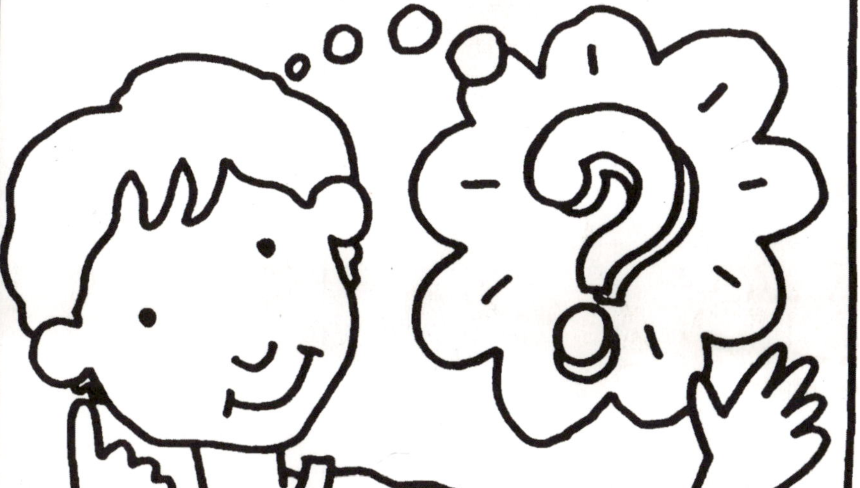

written by Katie John Sharp
illustrated by Jennifer Beck Harris

SAXON PUBLISHERS

Andrew

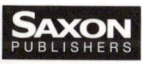

THIS BOOK IS THE PROPERTY OF:

STATE_____	Book No. _____
PROVINCE_____	Enter information
COUNTY_____	in spaces
PARISH_____	to the left as
SCHOOL DISTRICT_____	instructed
OTHER_____	

ISSUED TO	Year Used	CONDITION ISSUED	CONDITION RETURNED
....	....	....	....
....	....	....	....
....	....	....	....
....	....	....	....
....	....	....	....
....	....	....	....
....	....	....	....
....	....	....	....

PUPILS to whom this textbook is issued must not write on any page or mark any part of it in any way, consumable textbooks excepted.

1. Teachers should see that the pupil's name is clearly written in ink in the spaces above in every book issued.
2. The following terms should be used in recording the condition of the book: New; Good; Fair; Poor; Bad.